The Counting Zoo Crew:

0-10 Animal Numbers and Art

Written by Stacy J. Shaneyfelt
Illustrated by Mariia Riazantseva

Dedication

I dedicate this book to my sweet child, Lucena, an animal lover.
May you find miles of counting smiles within this book! XO

0 cows for our counting pals.

1 hedgehog loves to jog.

2 mice munch on rice.

Rat's
Takeout

3 butterflies wear bow ties.

4 snakes crave pancakes.

5 lions roar and race through the door.

6 snails wiggle their tails.

7 horses trot on golf courses.

8 beachy bears lounge in chairs.

zzZZ
Bees

9 raccoons hum cheesy tunes.

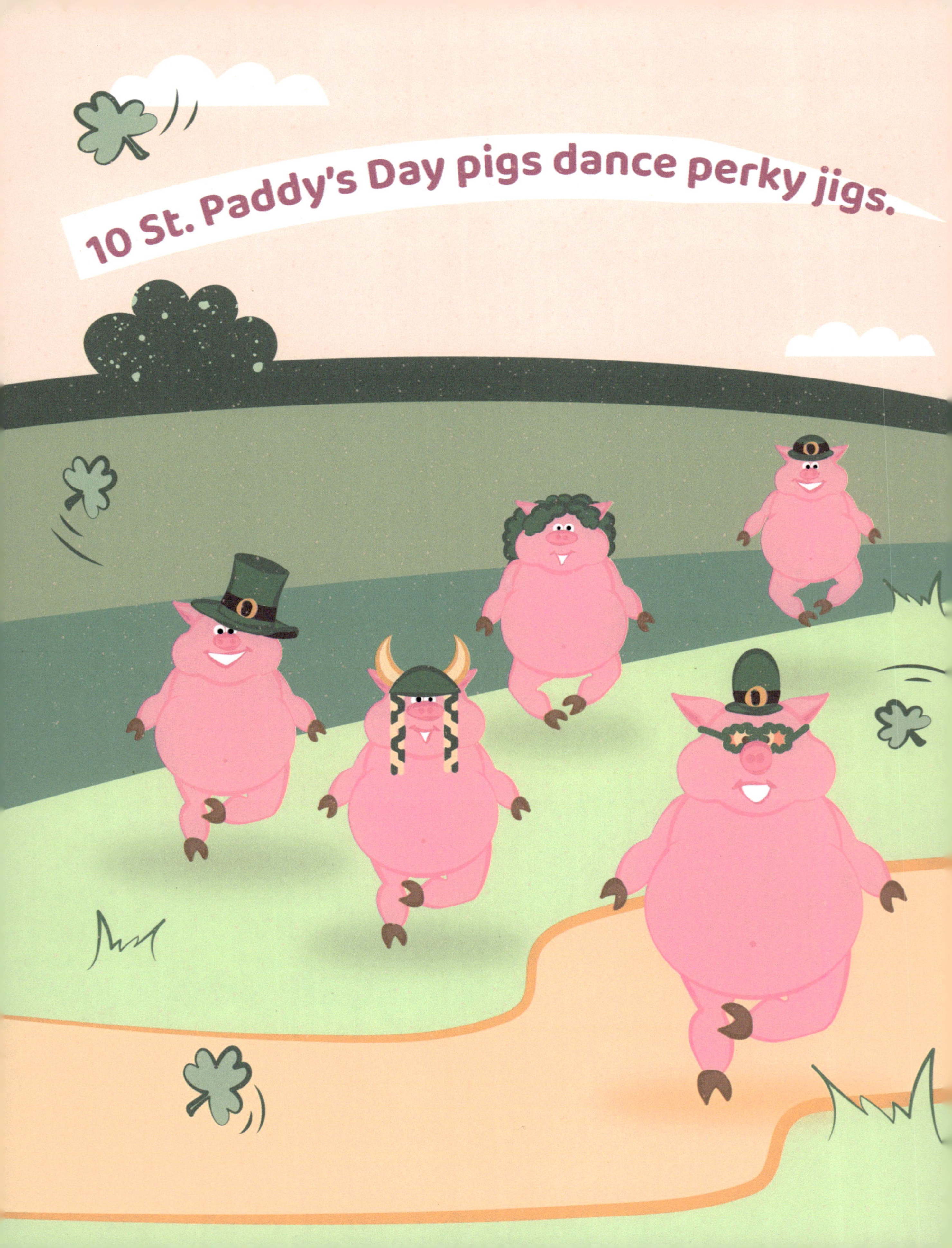

10 St. Paddy's Day pigs dance perky jigs.

ABOUT THE AUTHOR

After obtaining her BS in Secondary English Education and MA in English from Slippery Rock University of PA, Stacy embarked on a successful teaching career that spanned public, government, and charter schools in Pittsburgh, PA, Oklahoma City, Norman, OK, and Okinawa, Japan. Stacy proudly earned a 2004 Fulbright-Hays Seminar Scholarship to Thailand and Vietnam from the United States Department of Education, Teacher of the Year in two schools, as well as other teaching accolades.

In addition to multicultural and social activism, Stacy now savors sweet moments with her awesome husband, two fierce and fabulous daughters, and three frisky fur babies. Presently working as a virtual freelancer, private English and ESL tutor, and online editor/proofreader, she also enjoys films, travel, coffee, art, and all things mindful!

ABOUT THE ILLUSTRATOR

Mariia is currently studying motion-designer and illustrator in Russia.

Her digital illustrations is a tool for self-expression.

You can see her work at instagram @vfif_ria , UpWork https://www.upwork.com/freelancers/~01a5b9b5b5b5ee0d07.

Thank You Page

Thank you for buying this book. As a working mom and military spouse, your reviews mean so much to me because I aim to unite global readers through art and literacy. Kindly post a shor review on this book's Amazon page. I truly appreciate you for book buzzing with me!

If you like this book, then please check out my others at https://www.amazon.com/Stacy-Shaneyfelt/e/B08TVX7C5X/.

Coloring Sheet

Which Animal Is Absent?

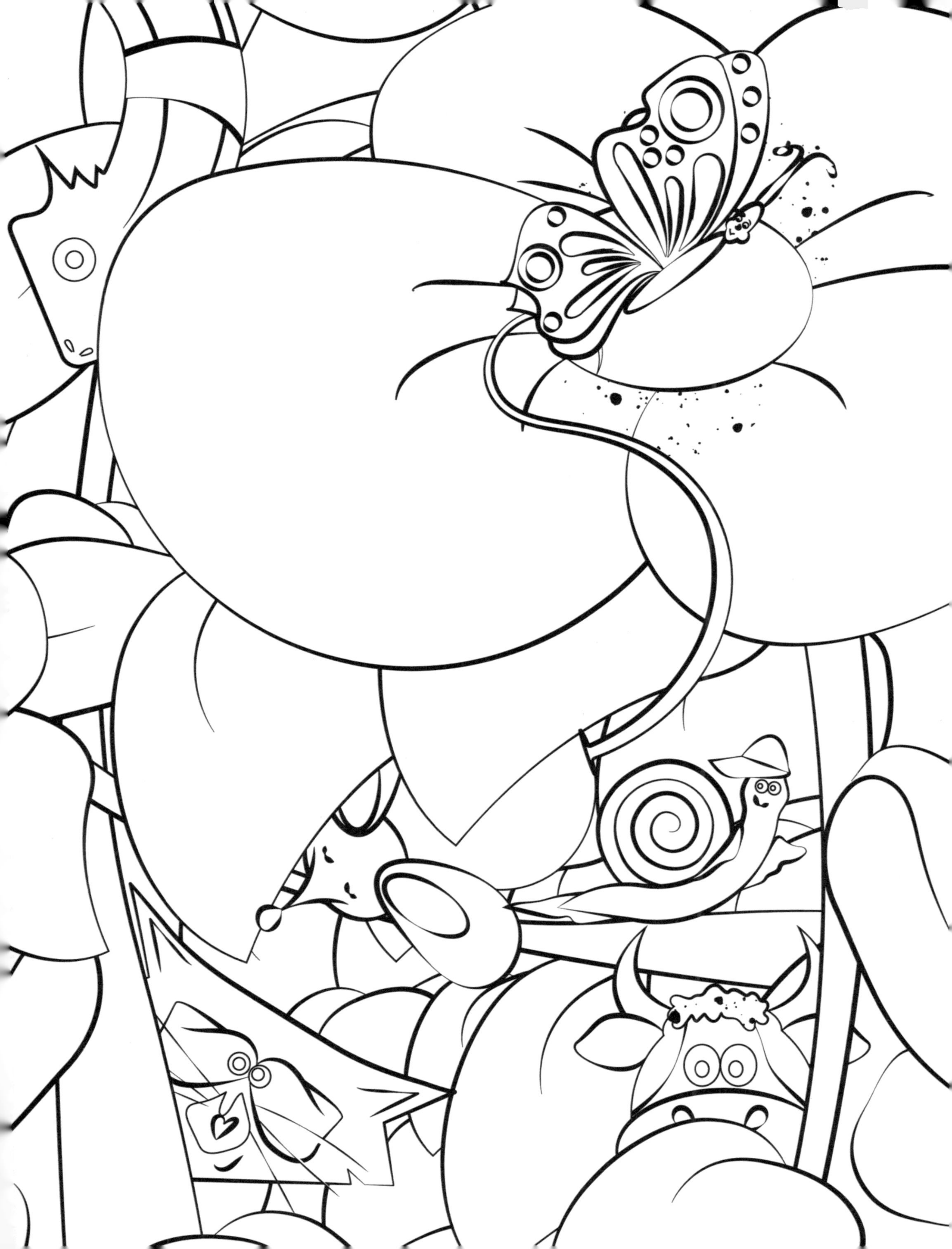

Made in the USA
Monee, IL
05 November 2026

67948394R00019